Water Buffalo

by Grace Hansen

Abdo Kids Jumbo is an Imprint of Abdo Kids
abdobooks.com

abdobooks.com

Published by Abdo Kids, a division of ABDO, P.O. Box 398166, Minneapolis, Minnesota 55439.
Copyright © 2021 by Abdo Consulting Group, Inc. International copyrights reserved in all countries.
No part of this book may be reproduced in any form without written permission from the publisher.
Abdo Kids Jumbo™ is a trademark and logo of Abdo Kids.

Printed in China

102020

012021

THIS BOOK CONTAINS
RECYCLED MATERIALS

Photo Credits: Alamy, iStock, Shutterstock

Production Contributors: Teddy Borth, Jennie Forsberg, Grace Hansen
Design Contributors: Dorothy Toth, Pakou Moua

Library of Congress Control Number: 2020910696
Publisher's Cataloging-in-Publication Data

Names: Hansen, Grace, author.

Title: Water buffalo / by Grace Hansen

Description: Minneapolis, Minnesota : Abdo Kids, 2021 | Series: Asian animals | Includes online resources
    and index.

Identifiers: ISBN 9781098205973 (lib. bdg.) | ISBN 9781098206536 (ebook) | ISBN 9781098206819
    (Read-to-Me ebook)

Subjects: LCSH: Water buffalo--Juvenile literature. | Asiatic buffaloes--Juvenile literature. | Animals--
    Juvenile literature. | Asia--Juvenile literature. | Endangered species--Juvenile literature

Classification: DDC 599.642--dc23

Table of Contents

Water Buffalo Habitat

The water buffalo belongs to the **bovine** family. It is the largest kind of bovine.

Water buffalo are **scattered** throughout Asia, in countries like India, Nepal, and Bhutan. They mainly live in grasslands and marshes. They are never far from water.

Asia
Nepal
Bhutan
India
N
W
E
S

Water buffalo spend much
of their day in muddy water.
This keeps them cool and
protects them from the sun.

9

Body

Water buffalo can stand more than 6 feet (1.8 m) tall at the shoulders. They can weigh up to 2,650 pounds (1,200 kg)!

Both males and females have

large, curved horns. But

a male's horns are longer,

growing to nearly 5 feet (1.5 m)!

Water buffalo have large, powerful bodies. They are dark in color and have some hair. Their tails are long and bushy at the tips.

Food

Water buffalo are plant eaters. They mainly feed on grass and **herbs**. But they will also eat plants found in water.

Herds & Babies

Water buffalo live in groups called herds. Each herd has around 30 members.

Females have calves every other year. Newborn calves weigh around 80 pounds (36 kg). Calves are protected by the herd. Young **bulls** will leave at 3 years old to form a new herd.

More Facts

* Water buffalo have wide hooves that keep them from sinking into muddy water.

* Water buffalo can live up to 25 years in the wild.

* Water buffalo have been domesticated for more than 5,000 years. They are very powerful and often used to plow fields.

Glossary

bovine – an animal of the cattle group, which also includes buffalo and bison.

bull – an adult male of several kinds of mammals, such as cattle and elephants.

herb – a flowering plant whose stem is soft rather than woody and that dies at the end of a growing season.

scattered – found in many locations.

23

Index